INFORMAL LOGICAL FALLACIES

A Brief Guide

Jacob E. Van Vleet

University Press of America,® Inc.
Lanham · Boulder · New York · Toronto · Plymouth, UK

4501 Forbes Boulevard
Suite 200
Lanham, Maryland 20706
UPA Acquisitions Department (301) 459-3366

Estover Road
Plymouth PL6 7PY
United Kingdom

Printed in the United States of America
British Library Cataloging in Publication Information Available

Library of Congress Control Number: 2010938998
ISBN: 978-0-7618-5432-6 (clothbound : alk. paper)
ISBN: 978-0-7618-5433-3 (paperback : alk. paper)
eISBN: 978-0-7618-5434-0

™ The paper used in this publication meets the minimum requirements of American National Standard for Information Sciences—Permanence of Paper for Printed Library Materials, ANSI Z39.48-1992

The age of making distinctions is gone.

Soren Kierkegaard

Contents

Preface

I wrote this book over a period of five years while teaching logic and critical thinking courses to undergraduate students in the San Francisco Bay Area. I found that logic textbooks created for these classes are usually large and inaccessible to the average reader. Furthermore, most of these textbooks spend a great deal of time covering formal logic rather than informal logic.

Informal Logical Fallacies: A Brief Guide was designed as an accessible supplementary text for logic or critical thinking courses. This book introduces the reader to over forty informal logical fallacies. These fallacies are commonly found in the media today and are often used to propagate fear, blind faith, and submission. For this reason alone, it is important to become aware of informal logical fallacies. It is my sincere hope that this short guide will sharpen the critical thinking skills of all who read it.

Jacob E. Van Vleet
Berkeley, California
July 2010

Introduction

WHAT IS AN INFORMAL LOGICAL FALLACY?

The word fallacy comes from the Latin word *fallacia*, meaning "trick," "deceit," or "fraud." A logical fallacy is an argument containing faulty reasoning. There are two types of logical fallacies: formal and informal. A formal logical fallacy is an argument that is flawed due to an error pertaining to the *structure* of the argument. An informal logical fallacy is an argument that is flawed due to an error pertaining to the *content* of the argument. This short guide is concerned with informal logical fallacies.

This book divides informal logical fallacies into five categories. First there are *linguistic fallacies*. These arguments are flawed due to a lack of clarity, as the use of vague or ambiguous terms severely weakens one's argument. Second, there are *fallacies of omission*. These arguments selectively leave out vital components or misrepresent certain positions in order to convince the listener of the correctness of the conclusion. For example, if certain options are left out of the equation, then the listener may assume that only the options presented are available. Third, there are *fallacies of intrusion*. These fallacies involve an incorporation of irrelevant material in order to persuade the listener to agree with a particular claim. For example, appeals to fear or pity may convince one to accept a conclusion when these appeals have little or nothing to do with the truth of the conclusion. Fourth, there are *fallacies involving built-in assumptions*. These arguments contain assumptions about tradition, nature, and other people. Fifth, there are *causal fallacies*. These arguments rest on a misunderstanding of cause and effect. For example, one might assume that simply because events occur sequentially they are causally related.

There are many kinds of logical fallacies within each of these five categories. Also, it is not uncommon to encounter certain fallacies that fit in more than one category. These will become clear as one reads through the explanations and examples in the following.

THE IMPORTANCE OF STUDYING INFORMAL LOGICAL FALLACIES

There are at least three main reasons for studying informal logical fallacies. First, by becoming aware of fallacies, one will be able to detect poor arguments.

This is an invaluably helpful skill that will guide one through a world full of spin, propaganda, and lies.

Second, by studying informal logical fallacies, one's analytical capabilities will increase. This means that one will not only be able to recognize erroneous reasoning but will also be able to think more critically in general. The ability to think critically will help one in every aspect of life.

Finally, by studying informal logical fallacies, one can gain the confidence to challenge lazy assumptions and unquestioned beliefs. This confidence is necessary in order to think for oneself. Only by thinking for oneself will one be able to fully and authentically approach the most important religious, political, and ethical issues of today.

Chapter One

Linguistic Fallacies

If you can't say it clearly, then you probably don't understand it yourself.

John Searle

1.1 FALLACY OF DIVISION

This is also known as the *whole-to-part fallacy.* This argument states that because the whole has a certain characteristic, the parts that make up the whole must also have the same characteristic. In other words, the fallacy of division claims that the whole is always the same as its parts. For example:

1. The Dodgers have an excellent team this year. Therefore, we can safely conclude that every player on the team is excellent.

2. Republicans are in favor of immigration reform. Mr. Thomas is a staunch Republican. Therefore, Mr. Thomas must be in favor of immigration reform.

3. Kay lives in the wealthy part of town. She must be wealthy.

4. Berkeley is a very liberal city. Stan lives in Berkeley. He must be an extremely liberal person.

5. The cake tastes very sweet. All of its ingredients must be sweet.

It is clear that the parts or ingredients of a whole do not necessarily have the same characteristics as the whole itself. When one commits the fallacy of division, one erroneously uses the characteristics of the whole to define or categorize its constituents. Such an argument strays far from clarity, making assumptions about the parts without critically investigating them or considering them individually.

1.2 FALLACY OF COMPOSITION

This fallacy is the opposite of the fallacy of division. It is sometimes referred to as the *part-to-whole fallacy*. The fallacy of composition involves an assumption that the characteristics of the parts are identical to the characteristics of the whole. In other words, the whole is assumed to be composed of identical parts. For example:

1. The woman who rear-ended me was Asian. All Asians must be terrible drivers.

2. Some living creatures exhibit signs of consciousness. Therefore, all living creatures must be conscious.

3. The cake recipe calls for salt. It must be a salty-tasting cake.

4. The terrorists who attacked us were Muslim. Therefore, all Muslims are terrorists.

5. My server at Sal's Diner was extremely inefficient. The service at Sal's is terrible and I will never go back.

If one ingredient of a recipe has a certain taste, this does not entail that the final product will taste like the ingredient. Likewise, simply because a member of a group or class is a certain way, this does not necessarily mean that the entire group or class is the same. Members of religions, political parties, and ethnic groups are all extremely diverse. However, due to this fallacy, stereotypical ways of thinking are common. We must be aware of stereotypes and other fallacies of composition.

1.3 VAGUENESS

Vagueness is perhaps the most common element found in unclear communication. A word or phrase is vague if it is imprecise or lacks specific clarifying details. For example:

1. It is obvious that Mary is special.

In this example the word "special" is not clear. Mary might be special because she is someone's lover, or because she is mentally challenged, or because she is entitled to some privilege. In any case, we cannot know because the statement is vague. Consider this example:

2. Person A: What do you do for a living?
Person B: I am an artist.

Here, we can see that the word "artist" is vague. Is the person a painter? A sculptor? A songwriter? A poet? Also, in this example we see another important aspect of vagueness. That is, people often offer vague responses in order to dodge questions. Consider these examples:

3. Judge: Where were you on the night of December fourteenth?
Defendant: I was out.

4. Veronica: How is your first semester going at UC Berkeley, George?
George: Oh, you know, it is what it is.

In both of these examples we see vague responses that are meant to dodge the question. In example 3 the defendant states that she was "out." It is unclear as to what this specifically means. In the next example George replies by stating "... it is what it is." This may appease Veronica; however, it is still a vague answer that does not reveal how George's first semester is going.

We must be on the lookout for ambiguous and vague words and phrases as we read the newspaper, watch the evening news, and talk with our friends. Also, we need to try and avoid vagueness and ambiguity ourselves, despite that it is often the easiest type of communication. Speaking and writing with clarity and precision takes work, but it ultimately reveals rather than obscures the truth.

1.4 EQUIVOCATION

This fallacy occurs when one word or phrase is used in two different ways in an argument. Sometimes the fallacy of equivocation is called the *double meaning fallacy*. Equivocation always relies on an ambiguous word or phrase and draws its conclusion from this ambiguity. For example:

1. All men are created equal. Women are not men. Therefore, women are not created equal.

In this example we can see that "men" is used quite differently in the first and second sentences. In the first, "men" refers to all human beings. In the second, "men" refers to only males. This argument is clearly fallacious because it relies on two very different definitions of "men."

Consider these examples:

2. Exercise is good for the heart. Johanna just broke up with her boyfriend and she said that she has a broken heart. She just needs to get some exercise.

3. Philosophers say that life is ultimately a mystery. My sister loves to curl up with a good mystery. She must truly enjoy life.

4. Everyone knows that they should do what is right. I have a right to smoke. Therefore, I should smoke.

In each of these examples we can see clearly that one word or phrase is used in two very different ways. We should always be aware of ambiguities and the use of double meanings. If an ambiguity or inconsistency is discovered in an argument, one should try and restate the argument in a clear manner, or one should dismiss it altogether.

1.5 WEAK ANALOGY

An analogy is a comparison of one thing with something else for the purposes of explaining or persuading. The comparison might be of two people, events, places, or just about anything else. For example, one might say, "Seattle is like a cleaner version of San Francisco." Here the comparison between two cities is clear and straightforward. However, some analogies are unclear, unhelpful, or downright nonsensical. These are called weak analogies. For example:

1. Renting a house, rather than buying, is like flushing your money down the toilet.

In reality, renting is not simply throwing away money as the argument suggests —even if it does not end in investment or ownership. Rather, renting is paying for the temporary use of property or a good. Many people use rented cars, houseboats, and movies among other things. Using something you've paid for is not the same as "flushing your money down the toilet."

Here are two more examples of weak analogies:

2. Denying the existence of God is like refusing to believe that the sky is blue.

3. Recycling is like putting a band-aid on a broken arm.

In both of these examples, the analogies given are unclear and unhelpful. Furthermore, the two things compared in each argument have nothing in common. We must strive to avoid using faulty analogies, and we must always be on the lookout for weak comparisons.

1.6 PSEUDO-PROFUNDITY

This fallacy is found when a statement appears to be insightful and profound, but in reality it is unclear and incoherent. We hear supposedly wise and weighty proclamations quite regularly from New Age gurus, televangelists, and college professors. But many of these statements, on closer inspection, are so vague that they might as well be meaningless. For example:

1. Guru: Reality does not exist; we exist on the shore of a collective nightmare.

> 2. Professor: My philosophy is that the ALL becomes the ONE which is identical to the Self and to the Cosmos through which knowledge is attained.

At first glance these statements appear intellectually deep. However, they lack clarity and specificity. Furthermore, each can be interpreted in a variety of ways. This can lead to multiple contradictory views, which in turn cause confusion.

The philosopher Martin Heidegger was often accused of committing the fallacy of pseudo-profundity. For example, philosopher Paul Edwards believes that Heidegger's philosophy is a perfect example of pseudo-profundity; or as Edwards puts it, "hideous gibberish."[1] Consider these excerpts of Heidegger's work:

> 3. Is Being a mere word and its meaning a vapor, or does what is designated by the word 'Being' hold within it the historical destiny of the West?[2]

> 4. We are ourselves the entities to be analyzed. The Being of any such entity is *in each case mine*. These entities, in their Being, comport themselves towards their Being. As entities with such Being, they are delivered over to their own Being. *Being* is that which is an issue for every such entity.[3]

These statements from Heidegger may be insightful or profoundly meaningful, but there is no way of knowing without further information from his writings. In any case, we must make it our goal to speak with clarity and coherence rather than vagueness and imprecision. Furthermore, we must listen critically to the statements of others, looking for clearness before assuming their words are profound or true—even when the speaker is an authority figure.

NOTES

1. Paul Edwards, "Heidegger's Quest for Being." In the Journal *Philosophy*, 1989. p. 468.

2. Martin Heidegger, *An Introduction to Metaphysics*. Translated by Ralph Mannheim. New Haven: Yale University Press, 1972. p. 42.

3. Martin Heidegger, *Being and Time*. Translated by John Macquarrie and Edward Robinson. New York: Harper Collins, 1962. p. 42.

Chapter Two

Fallacies of Omission

Anyone who doesn't take truth seriously in small matters cannot be trusted in large ones either.

Albert Einstein

2.1 BIFURCATION

The fallacy of bifurcation is also known as the *black-and-white fallacy*. This fallacy occurs when the arguer presents the listener with only two choices (hence the prefix *bi*), when in fact there are other possibilities to choose from. In other words, the argument limits the options of choices, omitting possible alternatives. The fallacy of bifurcation usually takes the following form:

Premise: There are only two options: x and y.

Conclusion: Option x is false, therefore option y must be true.

We can see clearly that, if there were indeed only two options, the argument would be valid. The fallacy occurs when the argument concerns factual human categories rather than merely logical symbols. For example:

1. You are either a Republican or a Democrat. You are not a Republican, so you must be a Democrat.

In this example only two political parties are presented in the premise, when in reality there are many others that should be considered: Green, Communist, Independent, etc. The fallacy of bifurcation can also be found in loaded questions or assertions, which, in themselves, are not arguments. For example:

2. Are you an agnostic or an atheist?

3. There are only two types of people in this country: those who love America and those who hate it.

4. You are either with us or against us.

In all of these examples the options are limited. This fallacy is commonly used by politicians in order to manipulate people. We often hear of "good versus evil" and "pro-war versus anti-war," when in fact there are other positions that need to be considered. The fallacy of bifurcation is always an oversimplification of a more complex issue.

2.2 FALSE DILEMMA

Like the fallacy of bifurcation, the false dilemma erroneously limits options. In fact, the bifurcation is a type of false dilemma. However, whereas the bifurcation limits the choices to only two, this fallacy limits the options to three or more when, in reality, there are other alternatives that should be considered. For example:

1. A menu reads: "We have meals for all diets: vegetarian, vegan, and meat eaters."

In this example, the options are limited to three, when in fact there are more types of diets than these. Some people follow religious dietary guidelines; some cannot eat gluten products; some only eat organic foods.

Here is a very different example of the false dilemma:

2. Jesus was either a liar, a lunatic, or Lord.

This example is a common argument for the divinity of Christ. However, this statement limits the choices, when there are other possibilities that should be considered. One alternate possibility, according to Muslims, is that Jesus was a prophet. Another alternative—albeit unlikely—is that Jesus never existed. In any case, all possibilities should be considered in an honest and open manner.

Here are some final examples of the false dilemma:

3. College students are either working toward an Associate's Degree, a Bachelor's Degree, or a Master's Degree.

4. Are you a size small, medium, or large?

5. Professor to students: Do you believe in Creationism, Evolutionary Theory, or Intelligent Design?

We should be aware of the false dilemma fallacy. One way to do this is by remembering that people, places, and events are often more complex than they seem.

2.3 ARGUMENTUM AD IGNORANTIAM

Argumentum ad ignorantiam is Latin for "argument from ignorance." This is an argument that uses a lack of proof or evidence to convince the listener that the subject in question cannot be taken seriously or should be discounted completely. Often the lack of proof or evidence is meant to imply a refutation.

The structure of this argument takes the following form:

> Premise: There is no evidence or proof for x.
>
> ---
>
> Conclusion: x is false; x should be dismissed; or, x has been refuted.

The argumentum ad ignorantiam fallacy is quite common in religious discussions. Consider this example:

> 1. The existence of God has never been disproved. Therefore, God must exist.

One of the serious problems with the ad ignorantiam fallacy is that in many cases its opposite can be argued in the same manner. For example:

> 2. The existence of God has never been proven. Therefore, God does not exist.

Simply because there is not verifiable evidence that a certain thing exists, this does not mean that its existence should be dismissed or not taken seriously. More importantly, we must remember that a lack of proof is not a refutation.

Here are three final examples:

> 3. Candidate Smith has never spoken out concerning her views on abortion. We can safely conclude that she must be pro-choice.
>
> 4. There is no proof that extraterrestrials exist. Therefore, they must be figments of our imagination.
>
> 5. Since we have no physical evidence supporting the claim that Socrates was a real person, he must have never existed.

Here we see a lack of evidence is used to dismiss the issue at hand. In some cases, this lack of proof is used to argue for the opposing position. When we encounter little or no evidence in support of an assertion or claim, we should be honest with ourselves instead of jumping to hasty conclusions. It is far better to admit that we do not have all the information needed to make an informed judgment, than to make arguments from ignorance.

2.4 SHIFTING THE BURDEN OF PROOF

Shifting the burden of proof is usually used as a response to a particular argument. It is a type of argumentum ad ignorantiam. This is an argument that maintains that because one cannot defend one's position, it must be incorrect or false. For example:

> 1. Lawyer: My client did not rob the bank last Friday. Prove that he did!

In this example, the lawyer rests his argument on the sole basis that his client cannot be proven guilty. However, as is the case with the argumentum ad ignorantiam, the lack of evidence in no way acts as a refutation of the argument's claim.

Consider these examples:

> 2. Brian: The CIA is in charge of cocaine distribution in the United States.
> Alex: That is crazy!
> Brian: Prove that it isn't so!

> 3. Preacher: God wants everyone to donate money to my ministry.
> Congregant: I don't think that's what God wants.
> Preacher: Well, provide evidence that God doesn't want it.

The burden of proof in each of these examples is shifted from the arguer onto the opponent. However, it is not the job of the opponent to disprove the arguer's position. It is always the arguer's task to provide solid evidence in support of his or her case. Without evidence, the arguer is simply making unsupported assertions.

2.5 STRAW ARGUMENT

The straw argument was originally named the *straw man argument*. It is an argument that intentionally presents a misrepresentation of a particular position in order to easily refute or dismiss it. A straw man is easier to knock down than a real person, hence the name. For example:

> 1. Jews and Christians simply believe that if you are good you will go to heaven, and if you are evil you will go to hell. This is the gist of their belief. So, you shouldn't take them seriously.

In this example the views of Jews and Christians are intentionally distorted so that they can be easily dismissed. However, one with an eye for fallacious argumentation will recognize that the argument rests on an oversimplification. The theological beliefs of Jews and Christians are far more complicated than what is stated in the argument.

Consider these examples:

2. Anarchism is a political ideology stating that everyone should run around and fornicate in the streets. It is for this reason that I reject anarchism.

3. The Democratic position is essentially a fascist and Communist stance. So, if you are in favor of Communism and fascism, then you will vote for the Democratic Party.

In both of these straw arguments, a misrepresentation is used in order to easily discredit or reject a position. We must be careful when we encounter one who claims to refute a position that he or she finds disagreeable. Often this is not a refutation of the actual position, but a refutation of a perversion of the position.

2.6 REDUCTIONISM

Reductionism occurs when one reduces a complex phenomenon to a single fact that may or may not be related to that phenomenon. This is often done in order to explain a many-sided or multifaceted issue, event, or belief system. Reductionism is regularly found in explanatory statements given for involved occurrences or systems of thought such as religion, politics, or science. Consider these examples:

1. Religion is simply a response to economic inequality.

2. Science is the ideology of modern academia.

3. Politics is nothing more than a power struggle.

Here we see the basic structure of the fallacy of reductionism: x (a complex issue) is nothing more than y (a simple, often vague, explanation). This is similar to the straw argument. The difference is this: the straw argument intentionally misrepresents a particular point of view in order to dismiss it, whereas reductionism often involves an over generalized explanation that precludes other explanatory factors. This may or may not be intentional, but like the straw argument, it is certainly intellectually lazy.

Reductionism is a clear example of a fallacy of omission. All fallacies of omission leave out important information rather than attempting to explore and discuss the various aspects of the issue at hand.

2.7 PERFECTIONIST FALLACY

This fallacy is committed when a person argues that a task or a goal should not be attempted because the possibility of reaching the goal or accomplishing the task is very unlikely. This argument is heard quite often when one rejects a certain business or political proposal due to the improbability of attaining it with perfection. For example:

1. We will never be able to fully eradicate crime on the streets of Oakland; therefore we should not waste our time and energy trying.

2. To create a coffee company that can truly compete with Java House is impossible. Therefore, we should not open a coffee shop.

In other words, the perfectionist fallacy states: if you cannot do something perfectly (or if you cannot finish something), then you should not even attempt it. This assumes that there is a *right way* to reach a goal, and that the end result is more important than the process or experience of working toward it. Another way of stating this fallacy is as follows: if you can't do x in a particular manner, then you shouldn't do it at all. Here are some final examples:

3. My artwork will never be as good as Picasso's or Chagall's. Therefore, I will not waste my time with it.

4. I keep going back to cigarettes. I have tried so many times to quit, but I can't. I'll always struggle with this. So, I guess I'll just keep smoking.

5. If you cannot be a perfect husband, then why do you want to get married in the first place?

6. I'll never be fluent in French, so I dropped my French class.

One of the main problems with this fallacy is its assumption that there is a universal definition of perfection, and that our actions should meet this definition and should be one hundred percent complete. In reality, many human goals or tasks get their value from the process of learning or making an effort rather than from their end results.

2.8 FALLACY OF THE UNKNOWABLE FACT

The fallacy of the unknowable fact contains as one of its premises a claim that is unknowable. This fallacy uses as evidence a "fact"—often speculation—that cannot be verified at the present time. For example:

1. Father to daughter: If you don't go to college, then you'll never be successful.

In this example, the father is claiming that he knows what the future holds for his daughter. However, the future is unknown, therefore there is no way to verify or falsify the father's claim.

Consider this example:

2. If only Adolf Hitler had a healthy relationship with his parents, World War II would have never occurred.

No matter what one finds out about Hitler's relationship with his parents, one cannot know how this relationship influenced his political decisions, let alone world history. To include this type of claim in one's argument is to include unverifiable data. In order to argue well, one must avoid appealing to the unknowable.

Here are two final examples of the fallacy of the unknowable fact:

3. I wish I'd gone to dental school instead of majoring in art. I'd be rich now!

4. Elizabeth should have married Stan instead of Wayne. She would be so much happier.

2.9 WILLED IGNORANCE

More of a diversionary tactic than a logical fallacy, willed ignorance remains a common response to rational argumentation. Willed ignorance occurs when one simply refuses to engage in logical discourse or critical thinking, maintaining that his or her beliefs about the issue are the final and absolute answer. This often occurs in scientific, political, or religious discussions. For example:

1. Person A: It seems to me that there are contradictions in the New Testament regarding the status and role of women.
Person B: Well, what the Bible says is true, and that's that—end of conversation!

2. Person A: There seem to be some inconsistencies in Einstein's theory of relativity.
Person B: Einstein was a genius, and what he said is indisputable, period.

3. Person A: After reading *The Republic*, I believe that Plato is endorsing fascism.
Person B: You have no idea what you are talking about, and I will not discuss Plato's work with you.

Willed ignorance is simply an inflexible, uncritical, unquestioning attitude that refuses to listen to the other side of an argument. A fundamental prerequisite for critical thinking is the ability to carefully and patiently consider various alternatives to the issue at hand. Critical thinking requires reflection and analysis of various viewpoints. This does not mean that all positions are correct—not at all! This simply means that one should avoid willed ignorance, and instead carefully analyze all arguments.

Chapter Three

Fallacies of Intrusion

The difficulty in philosophy is to say no more than we know.

Ludwig Wittgenstein

3.1 ARGUMENTUM AD HOMINEM, ABUSIVE

Argumentum ad hominem is translated as "argument against the person." The abusive form of this fallacy occurs when name-calling, slander or insults directed at a particular person are interjected into an argument. These are often used in order to manipulate the listener or audience. Usually, evidence or solid premises are substituted with personal attacks. Yet, on occasion, good premises are augmented with name-calling or derogatory terms. This is purely a psychological technique. It has nothing to do with solid reasoning or logical argumentation.

Here are some examples of the abusive ad hominem fallacy:

1. Candidate Greenstein is a bleeding-heart, unpatriotic liberal. Therefore, we cannot trust him as our leader.

2. Professor Searle is a real jerk! You should never take his class.

3. How can we trust a rabid Communist to teach our children about American politics?

One encounters this fallacy quite regularly in political campaigns and in radio talk shows. Even though the abusive ad hominem fallacy can be entertaining and even effective, it still relies on personal attacks rather than solid premises. It is for this reason that this argument should be avoided.

3.2 ARGUMENTUM AD HOMINEM, CIRCUMSTANTIAL

The ad hominem circumstantial fallacy, like the abusive version, seeks to attack the arguer rather than the argument. However, instead of resorting to personal attacks, an attempt is made to discredit the arguer's position by implying that he or she has impure or selfish motives. For example:

1. You should not listen to Professor Miller's arguments for faculty salary increases. The only reason he is arguing for a pay raise is because he himself will benefit from it.

In this example we see that, rather than his actual arguments being examined or critiqued, Professor Miller himself is being attacked. It claims that Miller has selfish motives for arguing in a certain manner, and because of this his arguments should be dismissed. The circumstantial fallacy does not only attack arguers and dismiss their specific arguments, as in the example above, but it can also be used to discredit any position or stance that one might hold. For example:

2. Harry was quite sure that Julie had only accepted his invitation to dinner because she felt sorry for him. This was probably the only reason she had been friendly with him in the first place.

3. It is obvious that Aunt Sue's Christmas gift to Emma was given with ulterior motives. She gave her an airline gift certificate, but she might as well have just asked Emma to come babysit.

4. I will not accept the administration's arguments for war. Clearly, the only reason the President wants to be in the Middle East is to secure natural resources.

In all of these examples, the arguments or actions alone are not taken seriously because certain circumstances have supposedly corrupted the source of each argument or action. We need to remember that not all arguments are presented with ulterior motives and that arguments should, if possible, be analyzed by themselves, apart from their sources.

3.3 ARGUMENTUM AD HOMINEM, POST MORTEM

This fallacy is committed when one argues that if a certain philosopher, scientist, or any other great thinker of the past were alive today, then he or she would engage in an ad hominem attack. This fallacy is heard quite often on college campuses—coming from students and professors alike. For example:

1. Person A: (Offers an interpretation or idea about an issue.)
Person B: (Responding) If Nietzsche were here, he would call you an idiot!

2. Person A: (States a critique of Plato's philosophy.)
Person B: (Responding) If Plato were here, he would say that your mind is not fit for learning and that you should simply be a laborer.

3. Person A: (Questions and objects to a Freudian idea.)
Person B: (Responding) Well, Freud would say that you are a sexually repressed jackass!

The post mortem ad hominem attack always uses an authority figure of the past (quite often someone who is dead, hence *post mortem*) to attack a person, idea, or argument. This fallacy is very easy to recognize. It almost always begins with, "If [person x] were here, he/she would say ..." This is fallacious because of the impossibility of being able to know what someone would say here and now, in our current historical and social context. We have no way of knowing what Plato, Nietzsche, or any other great thinker of the past might say if they were engaged in a philosophical discussion with someone at the present time. Even if we feel that we are an authority on a great thinker, this does not mean that we know what his/her exact response to an idea, issue, or argument would be today.

3.4 ARGUMENTUM AD VERECUNDIAM

Argumentum ad verecundiam is often translated from the Latin as "argument from authority." This fallacy relies on a false authority figure as support for a particular claim. A false authority is one who has little or no expertise in the particular context in which he/she is being used. The ad verecundiam argument can be found quite regularly in advertising. Often, celebrities, entertainers, or athletes will promote certain items or products of which they have no in-depth knowledge. For example, one might encounter a movie star selling pharmaceuticals, or a country singer promoting a politician.

A legitimate authority is one who is an expert in the appropriate field. For example, when a doctor states that a certain medication is needed, the patient should seriously consider this advice. Or, if a professor of mathematics states that an algebraic equation ought to be done in a certain way, the student should listen. If, however, one's doctor is preaching about religion, or if one's mathematics professor is pontificating about politics, both would be considered illegitimate, or at least questionable, authorities.

There are many other examples of the ad verecundiam fallacy. For example:

1. According to the bus driver, UC Berkeley is not a good university. So, you should reconsider attending in the Fall.

2. I know that democracy is the best form of government because my pastor said so.

3. My history professor gave a wonderful lecture on atheism. He is an expert, so he must be right.

In all of these examples false authorities are appealed to. When creating good arguments authorities may be used, however, they should be experts in the related field. Also, with this in mind, one must always remember that even experts can be wrong.

3.5 ARGUMENTUM AD BACULUM

In ancient Rome, a rod called a baculum was carried by each leader to symbolize his power. Named after this practice, the argumentum ad baculum fallacy is also known as the *argument from the rod,* or often it is simply referred to as the *argument from fear.*

The ad baculum argument uses threats, punishment or scare tactics to convince the listener to accept the conclusion of the argument. In short, fear replaces or is added to the premises or conclusion in order to psychologically manipulate the intended audience. The ad baculum fallacy usually takes the following form:

> Premise: Arguer presents his/her position, which includes threats or scare tactics.
>
> ---
>
> Conclusion: The arguer's position must be taken seriously or else harm may ensue.

Yet the ad baculum argument can also be found in many other forms. The common factor is always the appeal to fear. For example:

> 1. Sue, if you don't go out with me, remember that I know where your family lives.
>
> 2. If we don't take care of the terrorists, then our children and grandchildren will live in constant threat of being attacked.
>
> 3. No matter where you live, you are never safe from West Nile Virus. Therefore, you should buy the new West Nile Virus protective lotion available at your local drug store.

The appeal to fear can be a very powerful and effective form of argumentation. Politicians, preachers, and parents love to use ad baculum arguments. We must remember, however, that threats and scare tactics do not constitute solid premises.

3.6 APPEAL TO COMMON KNOWLEDGE

This fallacy occurs when solid premises are replaced with the fact that a belief is well known. It is sometimes referred to as the *every child knows fallacy* because it can include statements like, "even a child knows this," or, "this is so obvious that even a child would recognize it." Often the arguer will even appeal to "everyone" in his or her premises. However, the fact that something is widely accepted as true does not establish the truth of a conclusion. Here are some variations of this fallacy:

> 1. The earth is millions of years old. How do I know this? Because everyone knows this!
>
> 2. Every person is created equally. This is confirmed by the fact that everyone acknowledges this truth.
>
> 3. Even a small child can recognize morality and immorality. So, one cannot act as if any behavior is morally ambiguous!
>
> 4. Come on, everyone knows that Governor Wilkins is the best candidate for the position.
>
> 5. Of course you should accept the promotion! I can hardly believe you're asking for advice. We all know that money improves the quality of life.

As demonstrated in these examples, the arguer committing this fallacy uses an appeal to common knowledge rather than providing solid evidence for a claim. While this fallacy is heard everyday—it is indeed the backbone of peer pressure and the latest cultural trends—we must remember its inherent faulty reasoning.

3.7 GENETIC FALLACY

The genetic fallacy is an argument that assumes that a statement, position, or idea must be flawed if the source of the statement, position, or idea is flawed (or believed to be flawed). Often the genetic fallacy is an attempt to dismiss or refute another's position based solely on the origin of the position. The genetic fallacy usually has the following form:

> Premise: x originated from y.
>
> ______________________
>
> Conclusion: x must have similarities to y and therefore cannot be taken seriously.

This argument assumes that x is like y (presumably negative) simply because it came from y. This is flawed reasoning. Consider these examples of the genetic fallacy:

> 1. Why should we liberals not listen to Mr. Daniels? Because, Mr. Daniels is from a very conservative town in the Midwest.

2. I will not vote for Schroeder. Everyone knows she is German, and we all know German history. No anti-Semitic candidates for me!

In these examples, people are dismissed based on the erroneous assumption that they necessarily share certain negative (or perceived as negative) characteristics with their places of origin.

We must remember that a person or position should never be dismissed solely on the basis of origination. A position or argument should be judged on the basis of its strength or validity, not on its source.

3.8 ARGUMENTUM AD POPULUM

Argumentum ad populum, or *argument from the people*, is a fallacy that appeals to a group of people's beliefs, tastes or values. The people appealed to can be a small group, such as an individual clique or church group, or it can be a large group of people such as Republicans, Baptists, or feminists. The argumentum ad populum fallacy states that because a group holds a particular belief, attitude or value, this must lend support to the correctness of that belief, attitude or value. Also, the ad populum argument often claims that because the group is a certain way, the listener should accept the group's manner or action—or at least consider it as valid. For example:

1. The Southern Baptist Convention holds belief x. Therefore, you should hold belief x.

2. The Democratic Party endorses candidate Fuller for president. Because you are a registered Democrat, you should support candidate Fuller.

3. Our family has always acted in manner x. Since you are part of this family you should act in manner x as well.

In these examples, a group is appealed to as the authority on a particular issue. The fallacy occurs in assuming that the group is correct. Often the opinions of groups can be wrong. Consider outdated scientific and mythological beliefs. Consider forms of racism, sexism, and fascism practiced by many groups. Simply because a group of people hold a certain set of beliefs to be correct, this does not in any way imply the accurateness of these beliefs.

Argumentum ad populum is also known as *the democratic fallacy*. In a democracy, laws are passed and officials are given offices by the consensus of the largest number of people. However, simply because a large group of voters believe one course of action to be the best, this does not ensure that the vote of the majority is necessarily correct or right.

Beliefs, attitudes, and values need to be closely examined separately from the groups that support them. No matter who or how many people hold a certain belief, their advocacy does not prove its correctness.

3.9 APPEAL TO TRADITION

This fallacy is an argument that appeals to tradition or custom as justification for its conclusion. Traditions and customs can be admirable and are sometimes helpful when making decisions and formulating arguments. However, these alone do not provide solid premises for an argument. For example:

> 1. You should vote for a Republican. Your father and his father were both dedicated Republicans, so you should be too.
>
> 2. Mother to daughter: Our family has been Roman Catholic for many generations. How can you even consider leaving the church?
>
> 3. It has always been our custom to treat women this way. How can you challenge our customs?
>
> 4. Man: How can you be sure that your religious text is accurate?
> Priest: Because our tradition tells us that it is.

In each of these examples an appeal to tradition is found. The error in appealing solely to tradition is this: tradition by itself is not infallible. There have been many defective traditions (not to mention unjust and harmful ones). Simply because an idea or way of life has been a tradition, this does not guarantee that it is correct.

3.10 TU QUOQUE

Tu quoque is translated from the Latin as "you too." This fallacy is often used as a response to a particular claim, argument or suggestion. The tu quoque fallacy occurs when one charges another with hypocrisy or inconsistency in order to avoid taking the other's position seriously. For example:

> 1. Mother: You should stop smoking. It's harmful to your health.
> Daughter: Why should I listen to you? You started smoking when you were sixteen!

In this example, the daughter commits the tu quoque fallacy. She dismisses her mother's argument because she believes her mother is speaking in a hypocritical manner. While the mother may indeed be inconsistent, this does not invalidate her argument. Liars, manipulators, and even hypocritical parents can create good arguments. As with the genetic fallacy, one should not dismiss an argument based on its source. Consider these examples:

> 2. Father: You should not drink until you are twenty-one. It's unsafe and it's illegal.
> Son: How can I take you seriously? I know for a fact that you drank when you were under age!

3. Teacher: One should always recycle paper, plastic and glass products. It's good for the earth and it's good for future generations.
Student: Ha! I saw you throw your soda bottle in the trash after lunch! Why should we listen to someone who doesn't practice what she preaches?

As critical thinkers, we must consider the argument itself, separate from its source. This is often hard to do. In any case, the argument must be left to stand or fall on its own, regardless of hypocrisy or inconsistency on the part of the arguer.

3.11 TWO WRONGS MAKE A RIGHT

This fallacy assumes that because an unethical act has been committed, committing the same act again will somehow correct the initial unethical behavior. Simply repeating wrong behavior does not make it right. Like the tu quoque fallacy, the two wrongs make a right fallacy is often used to rationalize undesirable conduct. For example:

1. You took my sister's life, now I will take your sister's life.

2. Sharon: Why do you keep lying to your husband?
Amanda: Because he has been lying to me for years.

3. Jim: How can you live with yourself, knowing that you have killed five people?
Gene: The state kills people everyday; it's called the death penalty!

4. Alex: Aren't you going to return the wallet you found?
Harvey: Why should I? When I lost my wallet, no one returned mine!

In all of these examples an assumption is made that unethical behavior is somehow justified because others previously engaged in the same conduct. We must remember that this type of argumentation is seriously flawed. Mimicking another's actions does not justify one's own behavior.

3.12 RED HERRING

This fallacy gets its name from a practice once used in hunting. In order to hone their dogs' tracking skills, hunters would drag a herring across the hunting path, creating a distracting smell that might lead the dogs away from the correct path. The hunters would then train the dogs to avoid the distracting smell and to follow only the scent that they were instructed to follow. Like this practice, the red herring fallacy incorporates a distraction—an irrelevant issue—in order to divert attention away from the topic under discussion. For example:

1. Lawyer: Your honor, I know that my client murdered the bank teller. However, my client should not receive the maximum penalty. She was abandoned as a child, abused in foster homes, homeless by the age of ten, and she is currently addicted to drugs and alcohol. She is already suffering a great deal. She deserves mercy.

In this example, the lawyer incorporates irrelevant issues in order to divert the judge's attention away from the crime. These issues may be persuasive, however they are irrelevant to the topic at hand: murder. The hardships of the defendant's life do not change the fact that she murdered someone. (In addition to a red herring, this example is also an ad misericordiam fallacy.)

Consider these examples:

2. Police officer: Why do you have cocaine in your back pocket?
Man: I just got out of prison last night. I can't go back!
Police officer: Please answer my question.
Man: That is a nice wristwatch! Where did you get it?

3. Wife: I cannot believe that you forgot Valentine's Day again.
Husband: Did I tell you how beautiful you look in that dress?

4. Student: Professor, you stated that you believe our political system is profoundly flawed. Will you please explain this?
Professor: Your question reminds me of a trip I once took to Ecuador... (The professor goes on to tell a story without answering the question.)

In all of these examples an irrelevant issue is incorporated into the discussion in order to divert attention away from the topic at hand. This is precisely the goal of the red herring fallacy.

3.13 ARGUMENTUM AD CRUMENAM

Crumena is Latin for a purse that stores money. The argumentum ad crumenam assumes that wealth, money, or possessions are the measure of the correctness of an argument. For example:

1. We should carefully consider Senator Montgomery's political position. He was a very successful entrepreneur before he became a senator.

2. Ruth must be doing something right. Just look at all the nice clothes she has.

3. Bill Gates dropped out of college and now he is one of the wealthiest people alive. He must know what he is talking about.

In all of these examples it is assumed that a person's position is somehow more correct due to the wealth or possessions that he or she has. However, simply because one has a great deal of money, this does not ensure the validity of his or her stance on politics, relationships, or any other matter.

3.14 ARGUMENTUM AD MISERICORDIAM

The argumentum ad misericordiam is also known as the *appeal to pity*. It is a fallacy that appeals to emotions rather than providing solid evidence or facts to support its claim. By evoking pity, the arguer hopes to persuade the audience or listener to accept his or her conclusion. For example:

> 1. Boyfriend: You shouldn't break up with me.
> Girlfriend: Why?
> Boyfriend: Because the thought of losing you hurts so much.

In this example, the boyfriend is trying to evoke a sense of pity in his girlfriend so that she might stay with him. However, this is not a logical reason for staying with someone. Simply feeling sorry for another is no reason to agree with his or her claim.

Consider these examples:

> 2. Come on, I need you to lie for me. If you don't I will lose my job, my wife, and my kids.
>
> 3. Sue: Let's go to the philosophy lecture on Monday night.
> Brad: I cannot go; I have to do homework.
> Sue: If you don't go then I'll have to walk home alone in the dark. Are you sure you won't go?
>
> 4. Professor Dreyfus, please reconsider the grade you gave me on my final. If I don't get an "A," then I won't be able to get into graduate school.

In these examples, appeals to pity are made. Rather than providing good reasons in support of the claims, the arguer tries to arouse an emotional response. The ad misericordiam argument is often quite effective. However, we must be careful not to allow emotional responses to replace clear and coherent reasoning.

Chapter Four

Fallacies with Built-In Assumptions

To call an argument illogical, or a proposition false,
is a special kind of moral judgment.

C.S. Peirce

4.1 PETITIO PRINCIPII

This fallacy is also known as *begging the question.* From the Latin, petitio principii is translated as "appealing to the first principles." Petitio principii uses part of its conclusion in its own premise as support. This is a type of circular reasoning. Often this fallacy occurs when one uses two different words—one in the premises and one in the conclusion—which share the same definition. For example:

1. Abortion is wrong because abortion is unethical.

In this example, the conclusion (abortion is wrong) is supported by the premise (abortion is unethical). However, this argument begs the question because an act that is unethical, by definition, is an act that is wrong. Consider these examples:

2. I know that Gene and Susan slept together because I know that they had sex.

3. Your honor, the defendant should be held responsible for stealing my client's purse because he took something that did not belong to him.

4. The Bible is true because it contains no falsehoods.

5. Jan: Why did you decide to become a vegetarian, Anthony?
Anthony: Because I don't eat meat.

In these examples the conclusions are simply different versions of the premises. The conclusion should be established and supported by the premises; it should not simply be a restatement or variation of them.

4.2 COMPLEX QUESTION

The complex question is not an argument; it is a particular type of question. More specifically, it is a loaded question; i.e., a question with a built-in assumption. The complex question can often throw off the listener by forcing him or her to answer a flawed question. For example:

1. Lawyer to defendant: Where did you hide the murder weapon?

This example assumes that the defendant knows where the weapon is. If the defendant answers that he did not hide the murder weapon, then the judge and the jury might assume that he did have the weapon in his possession at some time. (In legal terms this is referred to as *leading the witness*.) The only way around this question is for the defendant to call to attention that the question is loaded.

Consider these examples:

2. When did you stop beating your wife?

3. When are you going to have children?

In example 2, the assumption is made that the person questioned did beat his wife. In example 3, the question assumes that the individual has decided to have children. The best response to these types of questions is not simply to answer the question as it has been posed. Rather, one should bring to light that the question has built-in assumptions, lest one fall into the trap set by the questioner. Consider the built-in assumptions in the following examples.

4. How many drinks have you had tonight?

5. Just how much did you enjoy the movie?

We must be careful to avoid asking loaded questions, and we must work to create straightforward, clear dialogue.

4.3 APPEAL TO HUMAN NATURE

This fallacy is an error in reasoning that appeals to "human nature" without clearly defining it. More specifically, it occurs when one uses this vague notion as justification for a particular human action. For example:

1. Well of course people worship celebrities—it's just our nature to worship them!

2. Defendant: Your honor, it's a dog-eat-dog world out there and I was doing what human nature is best at—defending and protecting myself! How can you hold someone responsible for his nature?

3. Humans by their very nature are political animals. Thus, it is clear that all individuals should participate in the political process.

It is important to remember that the appeal to human nature is often used as an excuse for certain behavior. The argument usually claims that one should not be held responsible because an action was simply in his or her "nature." One problem with this kind of appeal is that the notion of human nature itself is unclear. Many people speak of the nature of human beings, but on closer inspection it is difficult to specifically and clearly state what this means and what specific actions fall into this category. Furthermore, even if we were able to arrive at a universal definition, it would not necessarily mean that human nature could be used as a license for any type of behavior.

4.4 IS-OUGHT FALLACY

This fallacy occurs when an argument uses a purely descriptive statement (an "is" statement) as a premise, and moves to a prescriptive statement (an "ought" statement) as its conclusion. The words "is" and "ought" do not necessarily need to be in the argument for the is-ought fallacy to occur. There will, however, always be a description followed by a prescription. The is-ought fallacy has the following form:

Premise: Descriptive statement ("is").

Conclusion: Prescriptive statement ("ought").

Here are some examples:

1. Humans have carnivorous teeth. Therefore, you should eat meat.

2. It is on sale. Therefore you should buy it!

3. The Holy Scriptures say x. So, you should believe x.

As we can see from these examples, the is-ought fallacy usually is found when someone is trying to convince another person to act a certain way. However, we cannot assume that we can determine how things *ought* to be, just because we know how things *are*. To do so is flawed reasoning, because a prescriptive suggestion or command cannot be derived solely from a descriptive statement or observation. In other words, just because we have the facts, this does not necessarily mean we always know how others—or ourselves for that matter—should behave because of these facts.

It should be noted that this fallacy, unlike others, is debated by many philosophers. Some intellectuals, like John Searle at U.C. Berkeley, maintain the is-ought fallacy is not truly a fallacy, but a coherent and valid form of reasoning.[1] I will leave it to the reader to decide whether he or she believes the is-ought fallacy is indeed a fallacy or not.

4.5 PROJECTION

The fallacy of projection occurs when one "projects" personal feelings, beliefs, or attitudes onto another person without evidence to support this assumption. This is similar to the concept of projection found in certain psychological theories. Sigmund Freud believed that individuals often project unconscious desires and fears onto the external world. The fallacy of projection is different in that it is more generally applied to lazy assumptions that we all sometimes make about others. Projection can easily be described as assuming others' experiences of reality and/or conclusions concerning reality are the same as one's own. For example:

1. Of course Sharon will agree with us. She witnessed 9-11 also. Therefore, I am sure she will confirm our decision to go to war.

2. How can you possibly eat more than one piece of pizza? One piece always fills me up.

3. Jonathan is late to work again this morning. I am only late to work when I party the night before. Jonathan must have had a great time partying last night!

4. I can see God's handiwork clearly in nature. Anyone who looks at creation will surely come to the same conclusion.

In all of these examples one projects onto others one's personal beliefs, which have been derived from his or her individual experience. The problem with this type of reasoning is that individuals have different understandings of events, food, other people, nature, and of almost everything else in life. This does not mean that others' interpretations will always be different than one's own. In fact, many people share very similar experiences and interpretations. But, the key point to remember about the fallacy of projection is this: simply because one has arrived at a certain conclusion, it does not follow that others will make the same judgment.

4.6 NARCISSIST FALLACY

In Ancient Greek mythology Narcissus was a young man who was in love with his own image. Narcissism usually refers to a type of behavior that is egotistical or vain. The narcissist fallacy occurs when one assumes that others are thinking about him or her. This is similar to the fallacy of projection, where one projects one's beliefs, values, or feelings onto others. The key difference between these two fallacies, however, is that you commit the narcissist fallacy when you make the false assumption that people notice you, care about you, or are thinking about you. Consider these examples:

1. He smiled when I walked by. He must think these new jeans look really good on me.

2. My neighbor doesn't talk to me anymore. It's probably because he's jealous of my new car.

3. I'll go ahead and pull onto the freeway. Other cars will see me and let me in.

4. Today the mailman was rude to me and so was the toll booth attendant. I can't believe that people have been treating me so poorly on my birthday!

In each of these examples an assumption is made about others and their motivations and concerns. If we are honest with ourselves, we know that those who we encounter in everyday life are not necessarily thinking about us, even when they treat us in a certain manner. Other people have their own personal preoccupations—financial, marital, political, medical, spiritual, etc.—which often do not include us.

4.7 ANTHROPOMORPHISM

This fallacy gets its name from two Greek words: *anthropos*, meaning human, and *morphe*, meaning form. This fallacy occurs when one assumes that a non-human being or thing has the same characteristics as a human. Often this involves a projection of human emotions or behavior onto something non-human. For example:

1. Because of our lack of respect for nature, the earth has become angry with us humans.

2. As I look at the magnificent California redwoods, I feel them calling my name and extending their love to me.

3. When my pet snake Pepper smiles at me, I know she's satisfied.

4. I wish I had a better relationship with my computer. It keeps getting mad at me and freezing up!

In these examples, human emotions or attributes are assumed to be part of the non-human world. We must be honest with ourselves and try not to engage in projection of this sort. When we allow ourselves to make unfounded assumptions about nature and non-human creatures, we are reasoning with inauthenticity, laziness, and presumption. May our reasoning, instead, be built with the principles of critical thinking, which will help us avoid anthropomorphism and other fallacies.

4.8 SUBJECTIVIST FALLACY

Today it is quite common to hear the phrase, "That may be true for you, but it is not true for me." This is the quintessential subjectivist fallacy. This fallacy is called the subjectivist fallacy because many who use this erroneous type of argument mistakenly believe that all truth ultimately depends on the subject or the individual. This fallacy rests on a misunderstanding of what constitutes truth, confusing or conflating truth with belief.

It is critical that we distinguish between belief and truth. If something is true, then it describes accurately the way things are. As philosopher J.L. Mackie explains: "If we do take statements to be the primary bearers of truth, there seems to be a very simple answer to the question, what is it for them to be true: for a statement to be true is for things to be as they are stated to be."[2]

In contrast to truth, a belief is an opinion or conviction about something. A belief can be true or false. For example, if a person believes that 2 + 2 = 7, then he or she believes a falsehood, not a truth. Simply because one believes something, this does not make their belief accurate.

Also it is important to distinguish between truth and taste or preference. Someone might prefer Nietzsche's writings to Plato's. This is merely a matter of preference, not of truth. Or, someone might like the taste of tea over coffee. Again, this is simply one's taste; this is not an issue of truth or falsehood.

The subjectivist fallacy is quite commonly used as a mechanism for avoiding taking seriously the views and opinions of others. As critical thinkers, we should not take this easy method of escape. The beliefs and arguments of others should be considered in a serious and analytical manner, and they should be viewed from a standpoint that understands the vital difference between truth and belief.

NOTES

1. Searle, John. "How to Derive 'Ought' from 'Is,'" in *Philosophical Review*, 1964. 73: 43-58.

2. Mackie, J.L. *Truth, Probability and Paradox: Studies in Philosophical Logic*. Oxford: Clarendon Press, 1972. p. 22.

Chapter Five

Causal Fallacies

Nothing is so difficult as not deceiving oneself.

Ludwig Wittgenstein

5.1 FALLACY OF THE FALSE CAUSE

The fallacy of the false cause is an argument that assumes there is a causal relation between two events, without providing supporting evidence. Events often occur simultaneously or consecutively, however, this does not mean that they are necessarily related. For example:

1. The Chinese restaurant was robbed Friday night and again Saturday night. It must have been robbed by the same perpetrator.

2. Floods and tornadoes devastated the small town of Green Hills. It is clear that a Divine Being was punishing the town for its non-religious practices.

In the first example, the assumption is made that due to the same location of the robberies, the cause of both crimes must be the same. Without further supporting information, however, it is impossible to know if the same criminal was responsible for both events. In the second example, the assumption is made that two consecutive events (the town's non-religious practices and the floods and tornadoes) are causally related. But, like the first example, a causal relationship cannot be established.

All causal fallacies rest on a misunderstanding of cause and effect. The clearest examples of these are the fallacy of the false cause (above), the fallacy of the single cause, post hoc ergo propter hoc, and the slippery slope fallacy. These last three arguments will be explained next, followed by the gambler's fallacy and the appeal to consequences.

5.2 FALLACY OF THE SINGLE CAUSE

This fallacy is very similar to the fallacy of the false cause, described above. The fallacy of the single cause, however, states that an event has only one root or reason, when in fact there may be multiple causes. In contrast, the false cause fallacy is erroneous in its assumption of the cause in general. Consider this example of the fallacy of the single cause:

1. Mr. Rivers gunned down three innocent people because he was on drugs.

According to this claim, the cause for the shootings was drugs. However, situations like this are usually multifaceted and complicated. It is highly doubtful that drugs were the sole cause of Mr. Rivers' crime; drugs were most likely only one causal component of the complex event. Drugs combined with anger, fear, exhaustion, desperation, and/or ignorance might be a more accurate explanation. In any case, we simply do not know all of the causal factors involved in this case, so we need to refrain from assuming only one cause. Consider the following example:

2. The terrorists attacked us because they hate us.

In this argument, the stated conclusion is clearly an oversimplification. Hatred may or may not be one causal factor among several, but many other explanations should be considered when discussing terrorism. We must always remember that events often have more than one causal component.

5.3 POST HOC ERGO PROPTER HOC

Another fallacy of causation, the name of this fallacy translates from the Latin as "after this, therefore because of this." It is also known simply as the *post hoc fallacy*. This fallacy occurs when one presupposes that two consecutive but independently occurring events are causally related. In other words, one assumes that because one event happened after another event, the first event must have caused the second. For example:

1. Smith was elected Governor in May. In June, crime rates skyrocketed. Clearly, Smith must have caused crime to increase.

2. I lost my wallet the afternoon after I kicked my neighbor's cat. Therefore, my unethical act caused me somehow to lose my wallet.

3. Teen pregnancies have decreased since the creation of the Internet. So, it is clear that the Internet prevents teen pregnancies.

Like the fallacy of the false cause and the fallacy of the single cause, the post hoc fallacy rests on a misunderstanding of cause and effect. Remember, simply because two events occur sequentially, we cannot infer that the former event necessarily caused the latter. We must be careful not to project a causal relationship onto unrelated events.

5.4 SLIPPERY SLOPE

This fallacy occurs when one erroneously argues that if a certain act or event takes place, it will necessarily lead to a chain of events, ending in a dramatic or disastrous final result. The slippery slope is a fallacy precisely because we can never know if a whole series of events and/or a certain result is determined to follow one event or action in particular. Usually, but not always, the slippery slope argument is used as a fear tactic. Consider these examples:

> 1. You should not major in philosophy. If you do, you'll never get a good job, you'll lose friends, and no one will want to marry you. If you major in philosophy, you'll end up penniless and alone.
>
> 2. Playing the lottery will lead to gambling addiction, which means losing your house, your spouse, and everything that is precious to you. You'll find yourself homeless on the streets of Las Vegas, begging for change.
>
> 3. Marijuana is the gateway drug. Using it will lead to cocaine, heroine and crack use. Smoke a joint, and before you know it, you'll find yourself with an unquenchable desire for crack!

As we can see from these examples, the slippery slope fallacy claims to know the precise causal links and result following a certain event or action. Furthermore, these events usually end in a *negative result.*

Yet the slippery slope fallacy can also argue for a chain of causal events that necessarily lead to a *positive result.* For example:

> 4. Advertisement: Call now and purchase your own *Hollywood Complete Home Gym System*! Once it arrives, you'll be able to work out everyday, gaining energy, losing weight, and finally being in control of your life! By making the phone call today, you'll be strong and sexy in just a few months!

With this example, it becomes clear that the slippery slope can end in one of two extreme places: positive or negative. It should be noted, however, that the slippery slope always leads to a dramatic result, and always implies that the series of events leading to that result necessarily followed from the first action or event.

Sometimes, a slippery slope fallacy does not include or explain the linking events between the first and final ones. For example:

> 5. If Barbara rents her house to college students, the property value of the entire neighborhood will drop.

6. If we allow the government to listen in on our phone calls, then before we know it, even our sex lives will be monitored!

These two examples are slightly different versions of the slippery slope fallacy. They both follow the same pattern as presented above, but with the implicit assumption of negative causal events, rather than a clear presentation of a chain of events.

5.5 GAMBLER'S FALLACY

This fallacy of causation occurs when one mistakenly believes that an independently occurring event, such as a coin toss, has a causal effect on a separate event, such as a second coin toss. It is named the gambler's fallacy because this type of reasoning is commonly (but not exclusively) found in casinos. For example, if a person wins three hands of cards in a row, he or she might think, "I am on a roll! I am sure to continue winning!" However, there is not a causal relationship between previous card games and later card games—they are separate and non-causally related. Consider these examples:

1. I bought three lottery tickets, and I won money on the first two! I am sure to win on the third ticket, too!

2. The Raiders won the last two Monday nights. We can safely assume that they will win next Monday night as well.

As seen in the above examples, the gambler's fallacy is often used to predict that an upcoming event will be the same as a previous event. However, it is also commonly used to predict the opposite result. Here are some examples:

3. Our first two children were boys. This means that our next child will certainly be a girl.

4. I am sure that the Lakers will win their next game. How can I know this? Because they have lost five games in a row and they have to break their losing streak soon.

5. The last three guys I dated were losers. My next boyfriend can't possibly be a loser.

The fact that this fallacy can be used to predict opposite results makes its flawed nature glaringly evident. Yet the central issue here is this: the gambler's fallacy, like other causal fallacies, confuses the notion of cause and effect. One must remember that there are two types of events: causal and non-causal. A causal event is one in which an action or event necessarily causes a following event, such as a pool cue hitting a billiard ball, causing the ball to roll. A non-causal event is one that has no necessary or predictable following event, such as a winning raffle ticket foretelling future wins or losses.

5.6 APPEAL TO CONSEQUENCES

The appeal to consequences is a causal fallacy stating that the truth of a proposition is determined by the negative or positive consequences of that proposition. In other words, this fallacy is committed when one argues that a statement must be true or false, depending on whether the consequences are perceived to be good or bad. For example:

> 1. Hal: I know that God does not exist.
> Susan: What evidence do you have to support this claim?
> Hal: Well, just look at all of the harm that has been done in the name of God!

In this example, Hal does not provide solid premises to support his claim. Instead he points out that belief in God has had certain negative consequences. However, simply because harm has been done in the name of God, this does not mean that God does not exist. Likewise, simply because some churches and other religious institutions feed, clothe, and provide shelter for the poor, this does not necessarily mean that their religious beliefs are grounded in reality. Consider these examples:

> 2. Evolutionary theory is correct for many reasons. The strongest evidence for evolution lies in the fact that people who accept this theory are more civilized than those who do not.

> 3. I do not believe that the United States Military is presently killing innocent civilians in Iraq and Afghanistan. The thought of this is so disturbing and depressing that it cannot possibly be true.

In each of these examples the arguer appeals to the consequences of a particular belief in order to accept or dismiss it. The problem with this type of reasoning is that there is not a causal relationship between the entailments and the truth of a proposition. Consequences often have nothing to do with the truth or falsity of their cause.

Appendix I: Five Argument Forms

We have discussed many logical fallacies in this book. Now we will learn how to use and recognize good reasoning by introducing five traditional valid argument forms. For centuries, logicians and scientists have used these forms as templates for structuring valid arguments. These forms are: modus ponens, modus tollens, disjunctive syllogism, hypothetical syllogism, and constructive dilemma. Familiarity with these arguments is essential in the study of logic. They should be learned and remembered as prototypes of valid argumentation.

MODUS PONENS

Modus ponens is made up of two premises and a conclusion. It begins with a conditional statement as its first premise. A conditional statement is an "if-then" statement. For example: "If I drop the cup of coffee then it will spill." The "if" component is called the antecedent ("If I drop the cup of coffee ..."), and the "then" component is called the consequent ("... then it will spill"). The second premise of modus ponens always affirms the antecedent of the conditional statement. For example:

> If the printer prints in color, then it is out of black ink. (First premise)
> The printer is printing in color. (Second premise)
> Therefore, the printer is out of black ink. (Conclusion)

In this example, we see clearly that the antecedent of the first premise is affirmed in the second premise. In the conclusion, we see another characteristic always found in modus ponens: the consequent of the conditional premise is affirmed. In fact, modus ponens is translated from the Latin as the "mode of affirming." In symbolic form, modus ponens looks like this:

> If A then B.
> A.
> Therefore, B.

Modus ponens always follows this pattern. Here are some more examples of modus ponens.

1. If Bertrand Russell was a philosopher then he was an intellectual.
Bertrand Russell was a philosopher.
Therefore, Bertrand Russell was an intellectual.

2. If Mr. Jones murdered the bank teller in the botched robbery, then Mr. Jones is guilty and should be punished.
According to the video evidence, Mr. Jones did murder the bank teller in the botched robbery.
So, Mr. Jones is guilty and should be punished.

3. If I see the chef sneeze into the spaghetti, then I will not eat it.
I see the chef sneezing into the spaghetti.
I will not eat the spaghetti.

In all of these examples, one can clearly see the form of modus ponens. We should memorize this valid argument form and use it as a guide for structuring our own arguments.

MODUS TOLLENS

This valid argument form is similar to modus ponens. However, where modus ponens affirms the antecedent of the conditional statement, modus tollens denies the consequent of the conditional statement. This is why it is translated from the Latin as the "mode of removing." For example:

If the candle is lit, then there is a flame.
There is no flame.
Therefore, the candle is not lit.

Here, the consequent of the conditional statement is denied (or negated) in the second premise. And from this, it logically follows that the conclusion is true. Modus tollens takes the following symbolic form:

If A then B.
Not B.
Therefore, not A.

Here are some more examples of modus tollens:

1. If Steven was born in the United States, then he is a citizen of the United States.
Steven is not a citizen of the United States.
Therefore, Steven was not born in the United States.

2. If Sally has a cat, then she is a pet owner.
Sally is not a pet owner.
Therefore, Sally does not have a cat.

3. If the magician's assistant had been decapitated, then she would be dead.
The magician's assistant is not dead.
Therefore, the magician's assistant was not decapitated.

In these examples we see clearly the modus tollens pattern. Modus tollens, like modus ponens, is a valid and coherent logical form that should be remembered and employed.

DISJUNCTIVE SYLLOGISM

The disjunctive syllogism is perhaps the easiest valid form to remember. This form always has a disjunctive statement—an "either/or" statement—as one of its premises. For example:

Either Bach wrote Mass in B Minor or Wagner wrote it.
Wagner did not write Mass in B Minor.
Therefore, Bach wrote Mass in B Minor.

In the above, we are presented with two options in the first premise: Bach or Wagner. If these really are the only two choices, and if Wagner did not write Mass in B Minor, then the conclusion follows logically from the premises. If there were other choices that were intentionally left out, then this would be the fallacy of bifurcation.

Here we are reminded that all valid argument forms rest on the assumption that the premises are true. That is, the forms provide the structure for valid arguments, but in reality, the content of the forms may or may not be true. For the sake of learning the traditional argument forms and how to create good arguments, we will take the examples herein at face value. And, on this assumption, the above argument is valid.

The disjunctive syllogism has the following symbolic form:

Either A or B.
Not B (or, not A).
Therefore, A (or, B).

From this symbolic form we can clearly see that it does not matter which side of the disjunctive statement is denied. If one option is negated, then the other must be true. Consider these examples.

1. Either Holmes solved the crime or Dr. Watson solved the crime.
Dr. Watson did not solve the crime.
Therefore, Holmes solved the crime.

2. Either you go to college or you get a job.
You did not get a job.
So, you will go to college.

> 3. Truth is either a construct of institutions or it is the correspondence of language to reality.
> Truth is not a construct of institutions.
> Therefore, truth is the correspondence of language to reality.

The disjunctive syllogism is a valid form of reasoning when there are only two options to choose from. If possible, when creating good arguments, we should try to use forms like the disjunctive syllogism.

HYPOTHETICAL SYLLOGISM

The hypothetical syllogism is made up of three conditional statements: one in each of the two premises and one in the conclusion. This is also known as the *chain argument* because it links together all components of the premises and conclusion. For example:

> If I move to the Bay Area, then I will need an income.
> If I need an income, then I will need a job.
> So, if I move to the Bay Area, then I will need a job.

The symbolic form of the hypothetical syllogism is as follows:

> If A then B.
> If B then C.
> So, if A then C.

This argument form is quite easy to recognize because it consists entirely of conditional statements. Like the other valid argument forms, the hypothetical syllogism always provides an argument with good structure, but not necessarily true content. For the purpose of illustrating the hypothetical syllogism, we will assume the premises are true in the following three final examples of this form.

> 1. If community colleges continue to cut classes, then fewer Americans will have access to education.
> If fewer Americans have access to education, then poverty will increase.
> Therefore, if community colleges continue to cut classes, poverty will increase.

> 2. If I rob the bank, then I may kill someone.
> If I kill someone, then there is a possibility that I will spend my life in prison.
> So, if I rob the bank, then there is a possibility that I will spend my life in prison.

> 3. If the atheists are correct, then there is no God.
> If there is no God, then intelligent design is false.
> Therefore, if the atheists are correct, then intelligent design is false.

CONSTRUCTIVE DILEMMA

This is the final of the five valid argument forms we will discuss. The constructive dilemma uses both disjunctive and conditional statements. It also is the only valid argument form that uses three premises instead of two. Here is an example:

> We will either rent or buy a home.
> If we rent, then we will need to move.
> If we buy, then we will stay.
> So, we will either move or we will stay.

The symbolic form of this constructive dilemma is as follows:

> Either A or B.
> If A then C.
> If B then D.
> So, either C or D.

Here we see that the constructive dilemma begins with a disjunctive statement as its first premise. In the second and third premises, it contains conditional statements. Finally, it concludes with a disjunctive statement that affirms the consequents of the second and third premises. Here are some more examples of the constructive dilemma:

> 1. Either we celebrate Hanukkah or we celebrate Christmas.
> If we celebrate Hanukkah, then we will drive to Los Angeles.
> If we celebrate Christmas, then we will drive to San Francisco.
> Therefore, we will either drive to Los Angeles or to San Francisco.
>
> 2. Humans either have instrumental value or intrinsic value.
> If humans have instrumental value, then their worth is measured by their efficiency.
> If humans have intrinsic value, then they have worth regardless of how efficient they are.
> So, either human worth is measured by efficiency or humans have worth regardless of their efficiency.
>
> 3. Either Sigmund Freud was correct or C.S. Lewis was correct.
> If Sigmund Freud was correct, then the idea of God is simply a projection of unconscious desires.
> If C.S. Lewis was correct, then the theism of the Bible is true, and God exists independently of our unconscious desires.
> So, either God is a projection of our unconscious desires, or God exists independently of our unconscious desires.

In these examples, we clearly see the constructive dilemma employed. It is an especially helpful argument form to use when making plans in everyday life, or when debating abstract theories.

Like the other four valid argument forms we have discussed, the constructive dilemma is simply a template to be used when creating good arguments. Of course, it is not always possible to use one of these five forms. However, they are vital to the study of logic, and useful when one is seeking solid argument structure.

Appendix II: Exercises

PART ONE

Instructions: Create an argument that commits the fallacy listed below. If necessary, provide the context for each of your examples.

1. False dilemma

2. Slippery slope

3. Argumentum ad baculum

4. Argumentum ad hominem, abusive

5. Reductionism

6. Anthropomorphism

7. Tu quoque

8. Equivocation

9. Argumentum ad populum

10. Red herring

11. Shifting the burden of proof

12. Petitio principii

13. Appeal to consequences

14. Complex question

15. Fallacy of composition

PART TWO

Instructions: In the following, identify the informal logical fallacy or fallacies committed.

1. Marriage is just natural. It is a fundamental component of human life and it springs from human nature. To oppose marriage is to oppose human nature.

2. Plato was one of the first great philosophers. Plato taught that democracy was inherently flawed. It follows then, that all philosophers are opposed to democracy, and one should be cautious when reading their works.

3. The day after I broke up with my girlfriend, my car would not start. I am sure that she sabotaged my car engine.

4. Why am I supporting candidate Mulhauser for governor? Because Chuck Norris supports him! Chuck Norris is a very successful actor and he must know what he is talking about.

5. My dog is so much smarter than your cat. Ignorance is written all over your cat's face.

6. Seventy-three percent of all those surveyed outside the Bentonville Thrift-Mart opposed abortion. Therefore, it is safe to say that the majority of Americans oppose abortion.

7. I ate breakfast at the Greasy Spoon two years ago. The coffee was terrible. The Greasy Spoon is an awful restaurant.

8. Are you a college student or are you a drop-out?

9. Our new president is secretly a Socialist. How do I know this? Because my uncle Ricardo told me and my uncle is an honest man.

10. Boss: I need the report by Tuesday.
Employee: Why do you need it by Tuesday?
Boss: Because if I don't have it on Tuesday, you'll be out of work on Wednesday!

11. Christians say that God is love. Love is an emotion. So, God is simply an emotion.

12. Ken argues that the Iraq War is morally wrong. But, Ken is just bitter because his wife left him. He is just being emotional.

13. Keith: War is always caused by intolerant religious fanatics.
Owen: Really? What about the Civil War, World War I, World War II, the Korean War, and the Vietnam War?
Keith: Exactly! You have proven my point.

14. After the young man read Richard Dawkins' book *The God Delusion*, he took his own life. It is clear that Dawkins' book caused the man to commit suicide. Furthermore, Dawkins should be held responsible for the young man's death.

15. The Founding Fathers argued that all people are created equal. This is clearly untrue. Some people are better athletes than others; some are more intelligent than others. So, the Founding Fathers must have been wrong.

16. Either God created the world as stated in the book of Genesis, or life is a product of random chance. Both options cannot be true. If God created the world, then the Bible is true. If God did not create the world, then life is meaningless.

17. Bumper sticker: You can either save the trees or use plastic toilet paper!

PART THREE

Instructions: In the following, identify the informal logical fallacy or fallacies committed. Also, identify the type of fallacy committed, i.e., linguistic, omission, intrusion, built-in assumption, or causal.

1. It is perfectly ethical to experiment on animals. Animals are not human beings, so ethics do not apply to them.

2. Since we cannot come to a universal consensus concerning the definition of pornography, we should reject the entire discussion and move on to a more easily defined topic.

3. Communism is defined as the control of the masses by a fascist dictator who simply wants power and control. In Communist countries people have to wait in long lines for food and no one gets health care. Furthermore, the media is censored in those countries. For these reasons I reject Communism.

4. Laws against under age drinking are worthless. Teenagers are going to drink no matter what. It is time we get rid of under age drinking laws.

5. The only reason Professor Clark requires students to buy the book he wrote is so that he will make money.

6. I will not vote for a politician from the Bay Area. The last time I voted for a politician from the Bay Area, the economy collapsed!

7. Why should I listen to you when you tell me not to join the Army? You joined the Army when you were eighteen years old!

8. Lawyer: Why did you accept the bribe?
Politician: Well, everyone in Washington is a little corrupt.

9. Emma: I was abducted by aliens two years ago.
Kristopher: How do know that this is true? How can you be sure that it wasn't a hallucination or a dream?
Emma: Well, prove that it isn't true!

10. Professor to students: What do you think? Was Bill Clinton a great president or was he the greatest president?

11. There must be a universal, objective framework for ethics. If not, then everyone would just do whatever they pleased.

12. Crime is either caused by ignorance, a poor economy, or both. If ignorance is the cause, then a good education is the answer. If a poor economy is the cause, then Karl Marx was right and Capitalism should be replaced by a more just economic system.

13. I spent three years taking college classes and all I ended up with was a bunch of debt. I couldn't even get a decent job afterwards. College is a joke.

14. Nashmin: I believe that rape is wrong and it should carry a life sentence.
Randy: Rape has always been a part of human existence. We find examples of it in every society that has ever existed. Since it is a natural part of life, it shouldn't carry a life sentence.
Nashmin: That is not a good argument.
Randy: Well, neither one of us is really right. What is true for you isn't necessarily true for me.

15. Susan: Why should I vote for Senator Clark?
Paul: Because everyone knows that he is the best candidate.

16. My philosophy professor said that religion comes from a deep fear of not being able to explain our natural world. So, religion is bunk.

17. Love is simply a chemical reaction in the brain.

PART FOUR

Instructions: Identify the fallacy or fallacies committed in the following. If the example does not contain a fallacy, explain.

1. People who believe in God are like little children who believe in Santa Claus and the Tooth Fairy. When will they outgrow these childish beliefs?

2. I reject Senator Williams' proposal because he is an idiot!

3. Only Holmes and Watson investigated the crime. Either Holmes or Watson solved the crime. Holmes did not solve the crime. Therefore, Watson must have solved the crime.

4. We conclusively know that the Iraq War is immoral, because it is clearly unethical. And, anything that is unethical is certainly wrong.

5. Tests on Mr. Brown have not found any physical cause for his illness. Therefore, his illness must stem from a psychological disorder.

6. If we really believed in freedom and equal rights, then we would tear down all of the prisons in America. Inmates do not have freedom and equal rights.

7. The new car dealership occasionally has a good sale. If you go there to buy a car, they may or may not be having a sale.

8. It must be an excellent book. It has been on the New York Times Bestseller List for ten weeks!

9. Student: What should I do to succeed in your class?
Professor: The answer is simple: You need to get an A!

10. Detective: How do you know that Mr. Johnson is guilty?
Woman: I can't think of anyone else who would have committed the crime, so Mr. Johnson must have done it!

11. Studying symbolic logic is necessary in order to think clearly and coherently. Therefore, you ought to study symbolic logic.

12. Harvey: It is obvious that people have free will.
Bruno: How can you prove this?
Harvey: This is proven by the fact that people freely make choices everyday.

13. Dogs and cats are mammals. Max is a cat and Rex is a dog. Therefore, Max and Rex are mammals.

14. Ever since the rise of the Internet in the nineties, the number of marriages in the United States has gone down. It is clear that the Internet is damaging traditional marriage.

15. Punk rock music is dangerous because it causes young people to question authority figures. If young people question authority figures, then it will be impossible for a true democracy to exist.

16. Socrates said, "The unexamined life is not worth living." Joan, who works at the mini-mart, has obviously never examined her life. Therefore, her life is worthless.

17. People who convert to Catholicism just do so because they have an unconscious need for order and structure.

PART FIVE

Instructions: Answer the following questions about information presented in this book.

1. Which fallacy incorporates irrelevant information into the argument in order to distract?

2. Which fallacy presents the listener with only two options to choose from, when in fact there are more than two?

3. What type of argument assumes that the whole is identical with its parts?

4. What type of communication is often the easiest?

5. Which fallacy conflates truth with belief?

6. What type of argument contains an assumption that non-human things have human characteristics?

7. What types of fallacies involve the confusion of cause and effect?

8. Which argument simply restates its premise in a different way in its conclusion?

9. Which argument misrepresents another's position in order to dismiss it?

10. Which fallacy involves the assumption that a prescriptive statement can be derived from a purely descriptive statement?

11. Which fallacy incorporates name-calling or slander in order to manipulate the listener?

12. What attitude simply refuses to consider alternate possibilities and interpretations?

13. Which fallacy uses an appeal to pity?

14. Which fallacy is also known as the "democratic fallacy"? Why?

15. What type of argument incorporates threats or scare tactics?

16. Which fallacy involves a charge of hypocrisy?

17. Which fallacy occurs when a word that has more than one meaning is used in two different ways in an argument?

18. Which argument appeals to something unknowable as evidence for its conclusion?

19. Which fallacy appeals to money or wealth?

20. Which fallacy occurs when one assumes that two consecutive but independently occurring events are causally related?

PART SIX

Instructions: Identify the valid argument forms in the following examples (from *Appendix I: Five Argument Forms*).

1. If I am late to work then I will lose my job. I am late to work. Therefore, I will lose my job.

2. Either we win the game or we lose. If we win, then we stay in Atlanta. If we lose, then we go home to Chicago. So, either we stay in Atlanta or we go home to Chicago.

3. If the professor is late, then the class will be canceled. The class is not canceled. Therefore, the professor is not late.

4. If the unexamined life is not worth living, then I'd better start examining my life. If I start examining my life, then I'll need to question my beliefs. So, if the unexamined life is not worth living, then I'll need to question my beliefs.

5. Either empiricists are correct or rationalists are correct. If empiricists are correct, then all knowledge is acquired through the senses. If rationalists are correct, then some knowledge is innate. So, either all knowledge is acquired through the senses or some knowledge is innate.

6. If Aristotle was Plato's student, then he was familiar with Plato's philosophy. Aristotle was Plato's student. So, Aristotle was familiar with Plato's philosophy.

7. Either Hegel was influenced by Marx or Marx was influenced by Hegel. Hegel was not influenced by Marx. Therefore, Marx must have been influenced by Hegel.

8. If we remain complacent, then we will be safe. If we fight for what is right, then we be persecuted. Either we will remain complacent or we will fight for what is right. So, we will either be safe or we will be persecuted.

9. Understanding logic is necessary if one wants to recognize valid and invalid arguments. If understanding logic is necessary, then one will study it. Studying logic can be difficult, but if one wants to recognize valid arguments, then one will study logic.

10. If it is permissible for human beings to hunt animals for sport, then it is permissible for extraterrestrials to hunt human beings for sport. Humans have a long history of hunting animals for sport. It is permissible for humans to hunt animals for sport. Extraterrestrials do not hunt humans for sport, but it is permissible for them to do so.

11. Either Smith died from a gunshot wound or he died from a knife wound. It is not the case that Smith died from a knife wound. Therefore, despite the evidence, Smith must have died from a gunshot wound.

12. If we continue to consume oil at our present rate, then we will run out by the year 2040. If we run out of oil by the year 2040, then we will need to design new technology or begin to use technology that does not rely on oil. So, if we continue to consume oil at our present rate, then we will need to design new technology or begin to use technology that does not rely on oil.

PART SEVEN

Instructions: Identify the fallacy or fallacies committed in these quotations.

1. Men should either be treated generously or destroyed, because they take revenge for slight injuries—for heavy ones they cannot.[1]

Niccolo Machiavelli

2. There are no facts, only interpretations.[2]

Friedrich Nietzsche

3. Democracy never lasts long. It soon wastes, exhausts, and murders itself. There never was a Democracy that did not commit suicide.[3]

John Adams

4. Religion is the sigh of the oppressed creature, the heart of a heartless world, just as it is the spirit of spiritless conditions. It is the opium of the people.[4]

Karl Marx

5. All religions, with their gods, demigods, prophets, messiahs and saints are the product of the fancy and the credulity of men who have not yet reached the full development and complete possession of their intellectual development.[5]

Mikhail Bakunin

6. If nature does not wish that weaker individuals should mate with the stronger, she wishes even less that a superior race should intermingle with an inferior one; because in such a case all her efforts, throughout hundreds of thousands of years, to establish an evolutionary higher stage of being, may thus be rendered futile. History furnishes us with innumerable instances that prove this law.[6]

Adolf Hitler

7. Socialism is a philosophy of failure, the creed of ignorance, and the gospel of envy, its inherent virtue is the equal sharing of misery.[7]

Winston Churchill

8. What they have to discover, what all the efforts of capitalism's enemies are frantically aiming at hiding, is the fact that capitalism is not merely the 'practical,' but the only moral system in history.[8]

Ayn Rand

9. How do you tell a Communist? Well, it's someone who reads Marx and Lenin. And how do you tell an anti-Communist? It's someone who understands Marx and Lenin.[9]

Ronald Reagan

10. Indeed, I did have a relationship with Miss Lewinsky that was not appropriate. In fact, it was wrong. It constituted a critical lapse in judgment and a personal failure on my part for which I am solely and completely responsible. But I told the grand jury today and I say to you now that at no time did I ask anyone to lie, to hide or destroy evidence or to take any other unlawful action.[10]

Bill Clinton

11. Every nation in every region now has a decision to make. Either you are with us, or you are with the terrorists.[11]

George W. Bush

12. John McCain once opposed these tax cuts—he rightly called them unfair and fiscally irresponsible. But now he has done an about face and wants to make them permanent, just like he wants a permanent occupation in Iraq.[12]

Barack Obama

NOTES

1. Jay, Antony. (Editor) *Oxford Dictionary of Political Quotations*. Oxford: Oxford University Press, 2006. p. 250.

2. Ayer, A.J. & Jane O'Grady. (Editors) *A Dictionary of Philosophical Quotations*. Oxford: Blackwell, 1992. p. 318.

3. *Oxford Dictionary of Political Quotations*. p. 3.

4. *A Dictionary of Philosophical Quotations*. p. 285.

5. Bakunin, Mikhail. *God and State*. New York: Mother Earth Publishing Association. 1916. Chapter 2. From: www.marxists.org.

6. Hitler, Adolf. *Mein Kampf*. Translated by James Murphey. London: Hurst & Blackett, 1939. Chapter 11. From Project Gutenberg: www.gutenberg.org.

7. Churchill, Winston S. (Editor) *The Best of Winston Churchill's Speeches*. New York: Hyperion, 2003. p. 446.

8. Rand, Ayn. *Capitalism: The New Ideal*. New York: New American Library, 1966. p. 8.

9. *Remarks at the Annual Convention of Concerned Women for America*, Sept. 25, 1987.

10. *Oxford Dictionary of Political Quotations*. p. 97.

11. *Address to a Joint Session of Congress and the American People*, Sept. 20, 2001.

12. Quoted by Answini Anburajan in "Obama: McCain Represents a Third Bush Term." From: www.msnbc.com. March 20, 2008.

Appendix III: Characteristics of Critical Thinking

Critical thinking is defined as *the careful analysis of beliefs, arguments, and positions in order to arrive at correct conclusions.* By studying logical fallacies, one will become able to distinguish poor argumentation from good argumentation. However, familiarity with logical fallacies alone is not enough to turn one into a critical thinker. There are several other important characteristics of critical thinking. Here are some:

1. Openness to the possibility of being wrong.

2. Willingness to listen to others' arguments, ideas, and beliefs.

3. Ability to listen to criticism without responding emotionally.

4. Willingness to question one's own beliefs.

5. Commitment to making an effort to truly understand others' arguments, ideas, and beliefs before opposing or criticizing them.

6. Commitment to making an effort to avoid unclear language and arguments.

7. Willingness to respect alternate views if good support is provided for these views.

All of these characteristics of critical thinking require two things: courage and hard work. First, it is sometimes threatening to question one's own beliefs and to listen to others' arguments and ideas. This does not mean that critical thinking requires one to change his or her beliefs or to uncritically accept the beliefs of others. It simply means that one is willing to thoughtfully analyze both.

Second, critical thinking is not easy. It takes discipline and hard work to continually question, carefully consider, and test the many claims that we encounter. Many people avoid critical thinking for this very reason. It is far easier to allow someone else to think for you. However, if one wants to grow intellectually, then one will make an effort to think for oneself.

Glossary of Logic and Critical Thinking Terms

Absolutism: The belief that there are certain unchanging and absolute truths. See *subjectivism.*

Analogy, argument from: An argument that is based on a likeness between two people, places, events, acts or things. Usually arguments from analogy are used in order to prove that, due to a similarity between two things, these things must be similar in other respects as well.

Anecdotal evidence: Support or evidence for a claim based on limited personal experience.

Antecedent: The "if" component of a conditional statement. The antecedent is always followed by the "then" part of the statement. See *conditional statement.*

Argument: In philosophy an argument is not a disagreement. Rather, an argument is a claim with reasons that are given in support. If a claim has no support, it is not an argument (see *assertion*). An argument is made up of two components: premises and a conclusion. See *premise* and *conclusion.*

Assertion: An unsupported claim.

Assumption: Usually an implicit or unstated belief. In logic, a built-in assumption is an unstated belief that is presupposed by the premises. For example: "I know that God exists because the Bible states that he does." In this example, the premise "... because the Bible states that He does" contains a built-in assumption: that the Bible is correct.

Belief: A person's or group's opinion or conviction about something. Beliefs can be true or false. One can believe that 2 + 2 = 7; this does not mean that it is true. Beliefs can also concern particular tastes or preferences, which are not objectively true or false.

Causal event: An event that is causally related to a following event, such as a pool cue hitting a billiard ball. See *non-causal event.*

Conclusion: The claim of an argument which uses premises as support.

Conditional statement: An "if-then" statement made up of an antecedent and a consequent. For example: "If it rains, then the car will get wet."

Conjunctive statement: A statement that combines two distinct components with an "and." For example: "Dave's family is German *and* Lutheran."

Consequent: The "then" component of a conditional statement, which always follows the "if" part of the statement. See *conditional statement*.

Contradiction: A contradiction occurs when two statements cannot both be true. For example, if one states that she has never been to London and that she has been to London, both of these statements cannot be true.

Deductive argument: An argument in which the conclusion is *necessarily* true if the premises are true and if the conclusion follows coherently from them.

Disanalogy: An argument from analogy that is weak due to a lack of similarity between the things compared. Also known as a *weak analogy*.

Disjunctive statement: An "either-or" statement. For example: "*Either* Holmes solved the crime *or* Watson solved the crime."

Emotive language: Words or phrases specifically used in order to stimulate feelings.

Euphemism: A mild or inoffensive word or phrase substituted for a harsh or unpleasant fact or occurrence. For example, one might speak of "passing away" rather than dying.

Fact: Something that is indisputably true. For example, 2 + 3 = 5 is a fact, and it is also a fact that the earth is spherical.

Fallacy: An error in reasoning or faulty reasoning. There are two types of fallacies: formal and informal.

Falsification: To establish the falsity of a conclusion by introducing a counterexample that proves the conclusion false.

Formal fallacy: An argument that has a flaw in its structure, not in its content.

Hasty conclusion: A conclusion drawn from limited or insufficient evidence.

Inductive argument: An argument in which the truth of the conclusion is only *probable.*

Informal fallacy: An argument that contains a flaw in the content of the premises or conclusion. This often occurs due to built-in assumptions, omission or intrusion.

Informative language: Language that seeks to inform rather than to arouse feelings.

Invalid argument: A deductive argument in which the truth of the premises does not ensure the truth of the conclusion.

Law of non-contradiction: A basic law of logic stating that two statements cannot contradict each other and at the same time both be true. See *contradiction.*

Logic: The art or science of argumentation. Also, the study of methods used to arrive at correct conclusions.

Necessary condition: Any condition that, if absent, guarantees that an event will not occur. For example, a necessary condition for rain is the presence of clouds. (However, clouds do not guarantee that it will rain.) See *sufficient condition.*

Non-causal event: An event that does not necessarily have an effect, such as a coin toss. See *causal event.*

Non sequitur: Latin for "it does not follow." A non sequitur is any conclusion that does not follow logically from the premises.

Objective claim: A proposition that is not subjective. A truth claim that is not dependent on a specific individual's beliefs.

Philosophy: From the Greek, translated as the "love of wisdom." To practice philosophy is to think critically and to question beliefs, knowledge, truth and other important aspects of human life. Socrates (470-399 BCE) is said to be the "father" of Western philosophy. His most famous statement was: "The unexamined life is not worth living" (*Apology*, 38a). A philosopher is one who sincerely examines his or her life and life itself.

Premise: Facts or evidence from which a conclusion is derived. Premises are the foundation upon which a conclusion rests.

Proposition: The truth or falsehood expressed in a sentence.

Rationalization: The practice of making an excuse for one's behavior even when one knows he or she is in the wrong.

Reductio ad absurdum: From the Latin "reduction to the absurd." An argument that claims a statement should be dismissed if its consequences are illogical or "absurd."

Reason: 1. Evidence or premises in an argument. 2. The human capacity to think logically and rationally.

Rhetoric: The art of persuasive speaking or writing. The rhetorician's primary concern is effectively persuading an audience. This may mean he or she is willing to use fallacious arguments, emotive language and/or unsupported assertions in order to persuade.

Scapegoating: Occurs when one places the blame for a negative effect onto a person or group who are not responsible for it.

Semantics: The study of meaning in language. See *syntax*.

Signifier/Signified: A *signifier* is a word or linguistic term; *signified* is the mental concept that the signifier represents. For example, one might use the word "God" (signifier) to represent the idea of a transcendent, all-powerful being (signified).

Soundness: An argument is sound when it is valid and its premises are true.

Statement: A sentence that is either true or false.

Strong argument: An inductive argument in which is it probable, but not necessary, that the conclusion is true if indeed the premises are true.

Subjective claim: A proposition that is not objective. A truth claim that is dependent on a specific individual's beliefs.

Subjectivism: The belief that truth is entirely dependent on the subject or individual. See *absolutism*.

Sufficient condition: Any condition that, if present, guarantees that an event will occur. For example, decapitation of a human is a sufficient condition for death. In contrast, a gunshot is not a sufficient condition for death. One can be shot and still survive.

Syllogism: An argument made up of three components: two premises and a conclusion. The most famous example is:

Premise 1: All humans are mortal.
Premise 2: Socrates is a human.
Conclusion: Socrates is mortal.

Syntax: The rules and characteristics of sentence structure. See *semantics*.

Truth: In logic, a belief, statement, or proposition is true if it accurately describes how things actually are. See *subjectivist fallacy* on page 30.

Valid argument: An argument with valid structure or form. A deductive argument is valid if the truth of the premises logically ensures the truth of the conclusion. Validity is not a guarantor of truth, however. An argument may be valid and also contain false premises.

Verification: To establish the truth of a conclusion by empirical observation.

Weak argument: An inductive argument in which it is unlikely that the conclusion is true, even if the premises are true.

Recommended Reading

One of the best overall guides to informal logical fallacies is Douglas Soccio and Vincent Barry's *Practical Logic: An Antidote for Uncritical Thinking.* This textbook is a goldmine for those interested in fallacies and critical thinking in general.

The best introduction to formal, informal, and categorical logic is C. Stephen Layman's *The Power of Logic.* This textbook provides short, clear chapters on the basics of logic and critical thinking. The only negative features of this book, and of Soccio and Barry's, are the size: each is over 500 pages.

If one is looking for an inexpensive guide to critical thinking and logical terms, then one should turn to Nigel Warburton's *Thinking from A to Z.* This is a small book that can be thrown in a backpack and used for reference in the college classroom or anywhere else.

One of my favorite books on informal logical fallacies is Madsen Pirie's *The Book of the Fallacy: A Handbook for Intellectual Subversives.* This is out of print, but a new edition has been released under a new title: *How to Win Every Argument: The Use and Abuse of Logic.* This book provides a good overview of many informal logical fallacies.

A very helpful historical overview of the development of logic is found in William and Martha Kneale's *History of Logic.* This is a clear survey of the evolution of Western logic from Aristotle, through Scholasticism, and into the present era. Another useful history text is Anton Dumitriu's *History of Logic.* Dumitriu gives an insightful introduction not only to Western logic but also to Indian and Chinese logic.

Bibliography

Ayer, Alfred J. *Language, Truth and Logic*. New York: Dover, 1952.

Ayer, Alfred J. and Jane O'Grady. (Editors) *A Dictionary of Philosophical Quotations*. Oxford: Blackwell, 1992.

Baron, Jonathan. *Thinking and Deciding*. Third Edition. New York: Cambridge University Press, 2000.

Cassell's Latin Dictionary, compiled by D.P. Simpson. New York: Macmillan, 1977.

Carnap, Rudolph. *Logical Foundations of Probability*. Chicago: University of Chicago Press, 1950.

Copi, Irving M. *Introduction to Logic*. Sixth Edition. New York: Macmillan, 1986.

Detlefsen, Michael and David Charles McCarty and John B. Bacon. *Logic from A to Z*. London: Routledge, 1999.

Dumitriu, Anton. *History of Logic*. London: Routledge, 1977.

Edwards, Paul. "Heidegger's Quest for Being," in *Philosophy*, 1989. 64: 437- 470.

Feibleman, James K. *Assumptions of Grand Logics*. The Hague: Martinus Nijhoff, 1979.

Hamblin, C.L. *Fallacies*. London: Methuen & Co., 1970.

Heidegger, Martin. *Being and Time*. Translated by John Macquarrie and Edward Robinson. New York: Harper, 1962.

———. *An Introduction to Metaphysics*. Translated by Ralph Mannheim. New Haven: Yale University Press, 1972.

Herrick, Paul. *The Many Worlds of Logic*. Second Edition. Oxford: Oxford University Press, 2000.

Hume, David. *An Enquiry Concerning Human Understanding*. Cambridge: Hackett, 1977.

Jay, Antony. (Editor) *The Oxford Dictionary of Political Quotations*. Oxford: Oxford University Press, 2006.

Kneale, William and Martha Kneale. *The Development of Logic*. Oxford: Clarendon Press, 1985.

Layman, C. Stephen. *The Power of Logic*. Third Edition. New York: McGraw-Hill, 2005.

McInerny, D.Q. *Being Logical: A Guide to Good Thinking*. New York: Random House, 2005.

Moore, Brooke Noel and Richard Parker. *Critical Thinking*. Eighth Edition. New York: McGraw-Hill, 2007.

Pirie, Madsen. *How to Win Every Argument: The Use and Abuse of Logic*. London: Continuum, 2006.

———. *The Book of the Fallacy: A Handbook for Intellectual Subversives*. London: Routledge, 1970.

Rolf, Bertil. "A Theory of Vagueness," in *Journal of Philosophical Logic*, 1980. 9: 315-325.

Russell, Bertrand. *Problems in Philosophy*. Cambridge: Hackett, 1972.

Searle, John. "How to Derive 'Ought' from 'Is,'" in *Philosophical Review*, 1964. 73: 43-58.

Soccio, Douglas J. & Vincent E. Barry. *Practical Logic: An Antidote for Uncritical Thinking*. Fifth Edition. Belmont: Wadsworth, 1998.

Sorenson, Roy A. *Blindspots*. Oxford: Clarendon Press, 1988.

Van Vleet, Jacob E. *Informal Logical Fallacies: A Brief Guide*. Berkeley: Independent Scholar's Press, 2009.

Walton, Douglas H. *Informal Logic: A Handbook for Critical Argumentation*. Cambridge: Cambridge University Press, 1989.

Warburton, Nigel. *Thinking from A to Z*. Third Edition. London: Routledge, 2007.

Weston, Anthony. *A Rulebook for Arguments*. Fourth Edition. Cambridge: Hackett, 2008.

Wittgenstein, Ludwig. *Tractatus Logico-Philosophicus*. Translated by D.F. Pears and B.F. McGuinness. London: Routledge, 1974.

———. *Philosophical Investigations*. Translated by G.E.M. Anscombe. Oxford: Blackwell, 1973.

Index

About the Author

Jacob Van Vleet teaches Philosophy at Diablo Valley College in Pleasant Hill, California. He is also adjunct Professor of Philosophy and Religious Studies at Ohlone College. He lives in Berkeley with his wife, Moriah.